CONTENTS

Boldface words appear in Words to Know.

AT THE FIREHOUSE

Firehouses are the buildings where firefighters work. Some places have **career** firefighters. Many other firefighters are **volunteers**. Firefighters may eat and sleep at the firehouse during their long **shifts**. They wait to be called to a fire or other **emergency**.

Places in My Community

DISCOVER!

Visiting the Firehouse

By Seth Lynch

Enslow Publishing

Please visit our website, www.enslow.com.
For a free color catalog of all our high-quality books, call toll free 1-800-398-2504 or fax 1-877-980-4454.

Cataloging-in-Publication Data

Names: Lynch, Seth.
Title: Visiting the firehouse / Seth Lynch.
Description: New York : Enslow Publishing, 2023. | Series: Places in my community | Includes glossary and index.
Identifiers: ISBN 9781978521261 (pbk.) | ISBN 9781978521285 (library bound) | ISBN 9781978521292 (ebook)
Subjects: LCSH: Fire stations–Juvenile literature. | Fire fighters–Juvenile literature..
Classification: LCC TH9148.L96 2023 | DDC 628.92–dc23

Published in 2023 by
Enslow Publishing
2544 Clinton Street
Buffalo, NY 14224

Designer: Andrea Davison-Bartolotta
Editor: Kristen Nelson

Photo credits: Cover (main) The Image Party/Shutterstock.com; cover (cartoon firetruck), back cover, pp. 6, 20 (firestation, firetruck) cloverlittleworld/Shutterstock.com; series art (town background) monbibi/Shutterstock.com; p. 4 Luis Louro/Shutterstock.com; p. 5 Ivelin Denev/Shutterstock.com; p. 6 (firefighters) mangsaabguru/Shutterstock.com; p. 7 Tyler Olson/Shutterstock.com; p. 9 ChameleonsEye/Shutterstock.com; p. 10 vectorpouch/Shutterstock.com; p. 11 dlewis33/istockphoto.com; p. 12 Oleksa13/Shutterstock.com; p. 13 Simone Hogan/Shutterstock.com; p. 14 Andrew Rybalko/Shutterstock.com; pp. 15, 17 (inset) Sergey Mironov/Shutterstock.com; p. 16 (dog) Oliver Hoffmann/Shutterstock.com; p. 16 (hydrant) meowu/Shutterstock.com; p. 17 (main) Julian Popov/Shutterstock.com; p. 18 GOLFX/Shutterstock.com; p. 19 Dan Holm/Shutterstock.com; p. 21 Manuel Esteban/Shutterstock.com.

Printed in the United States of America

CPSIA compliance information: Batch #CWENS23: For further information contact Enslow Publishing at 1-800-398-2504.

Firefighters are working at the firehouse all day and night!

WHEN TO VISIT

People in the community and school groups are commonly asked to set up a time to visit. This lets those on **duty** at the firehouse know who is coming and when. Firefighters show groups around and talk about life at the firehouse.

You may visit a firehouse with your family or your school.

WHAT TO EXPECT

Firehouse visits are not long. Firefighters don't have much time to spare. They keep the firehouse clean. They fix fire trucks and firefighting gear. Firefighters plan for future emergencies. They spend time training or learning better ways to do their job.

Firefighters also teach the community about fire safety.

Some firehouse tours will take visitors to see where the firefighters live when they are not on a **call**. Firehouses have rooms with TVs, beds, and computers too. Firefighters take turns sleeping during a long shift.

Firehouses have a kitchen where the firefighters make their meals.

SEE THE GEAR

Fire trucks are the highlight of any firehouse visit! You may have a chance to sit in the driver's seat or even turn on the **siren**. Sometimes visitors get to try out the hoses! Firehouses might have an **ambulance** you can tour too.

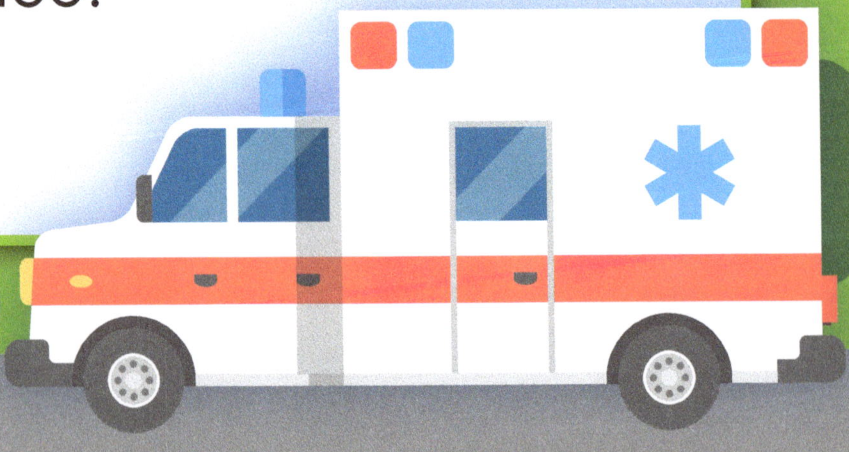

Firehouses may have a few different kinds of trucks.

Every firehouse tour is different. Visitors may have a chance to try on firefighting gear. They may see a firefighter slide down the fire pole in the building! Other firehouses might have firefighting items from history to look at.

Firefighters wear special clothes to stay safe during emergencies.

FIREHOUSE DOGS

Some firehouses have a dog! Many years ago, these dogs were guards. Today, most firehouse dogs are there to cheer up the firefighters on duty. They may be cute, but it's important to ask before petting a firehouse dog.

The spotted dalmatian is a common dog found at firehouses.

TIME TO LEARN

Visiting the firehouse can save your life! Firefighters often give visitors fire safety tips. You can ask questions too. You can ask about the best way to call for help and how to stay safe during emergencies.

Be sure to thank any firefighters that take the time to talk with you!

A QUICK EXIT

It's important to remember that firehouses are the home base of on-duty firefighters. A firehouse visit can end at any moment—even if it was planned. Firefighters may be called to a car **accident** or a house fire during your tour.

If a call comes in, visitors should leave quickly and quietly.

WORDS TO KNOW

accident: A sudden event that happens by chance and causes harm.

ambulance: A kind of truck or van that is used to take sick or hurt people to the hospital.

call: A request for someone or a team to do something.

career: A job someone chooses to do for a long time.

duty: Working at a certain time.

emergency: An unexpected situation that needs quick action.

shift: The period of time during which a person works.

siren: A piece of equipment that produces a loud, high-pitched warning sound.

volunteer: Someone who does a job without being paid.

FOR MORE INFORMATION

BOOKS

Gerry, Lisa, and Lori Epstein. *Dolley the Fire Dog*. Washington, DC: National Geographic, 2019.

Honders, Christine. *What's It Really Like to Be a Firefighter?* New York, NY: PowerKids Press, 2020.

WEBSITES

Fire Safe Kid – Kids Zone
www.firesafekid.org/kids-zone-2/
Check out this site to learn more about staying safe if there is a fire.

National Geographic Kids
juniors.nvfc.org/
Find out more about how to become a junior firefighter.

Publisher's note to educators and parents: Our editors have carefully reviewed these websites to ensure that they are suitable for students. Many websites change frequently, however, and we cannot guarantee that a site's future contents will continue to meet our high standards of quality and educational value. Be advised that students should be closely supervised whenever they access the internet.

INDEX